CHAPTER 1
ELECTRICAL CONDUCTORS

LEARNING OBJECTIVES

Learning objectives are stated at the beginning of each chapter. These learning objectives serve as a preview of the information you are expected to learn in the chapter. The comprehensive check questions are based on the objectives. By successfully completing the OCC-ECC, you indicate that you have met the objectives and have learned the information. The learning objectives are listed below.

Upon completing this chapter, you should be able to:

1. Recall the definitions of unit size, mil-foot, square mil, and circular mil and the mathematical equations and calculations for each.

2. Define specific resistance and recall the three factors used to calculate it in ohms.

3. Describe the proper use of the American Wire Gauge when making wire measurements.

4. Recall the factors required in selecting proper size wire.

5. State the advantages and disadvantages of copper or aluminum as conductors.

6. Define insulation resistance and dielectric strength including how the dielectric strength of an insulator is determined.

7. Identify the safety precautions to be taken when working with insulating materials.

8. Recall the most common insulators used for extremely high voltages.

9. State the type of conductor protection normally used for shipboard wiring.

10. Recall the design and use of coaxial cable.

ELECTRICAL CONDUCTORS

In the previous modules of this training series, you have learned about various circuit components. These components provide the majority of the operating characteristics of any electrical circuit. They are useless, however, if they are not connected together. Conductors are the means used to tie these components together.

Many factors determine the type of electrical conductor used to connect components. Some of these factors are the physical size of the conductor, its composition, and its electrical characteristics. Other factors that can determine the choice of a conductor are the weight, the cost, and the environment where the conductor will be used.

CONDUCTOR SIZES

To compare the resistance and size of one conductor with that of another, we need to establish a standard or unit size. A convenient unit of measurement of the diameter of a conductor is the mil (0.001, or one-thousandth of an inch). A convenient unit of conductor length is the foot. The standard unit of size in most cases is the MIL-FOOT. A wire will have a unit size if it has a diameter of 1 mil and a length of 1 foot.

SQUARE MIL

The square mil is a unit of measurement used to determine the cross-sectional area of a square or rectangular conductor (views A and B of figure 1-1). A square mil is defined as the area of a square, the sides of which are each 1 mil. To obtain the cross-sectional area of a square conductor, multiply the dimension of any side of the square by itself. For example, assume that you have a square conductor with a side dimension of 3 mils. Multiply 3 mils by itself (3 mils × 3 mils). This gives you a cross-sectional area of 9 square mils.

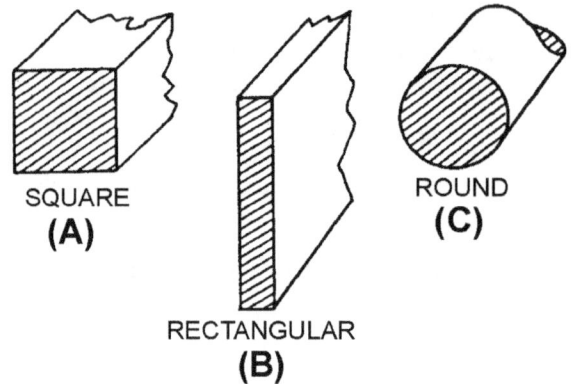

Figure 1-1.—Cross-sectional areas of conductors.

Q1.   State the reason for the establishment of a "unit size" for conductors.

Q2.   Calculate the diameter in MILS of a conductor that has a diameter of 0.375 inch.

Q3.   Define a mil-foot.

To determine the cross-sectional area of a rectangular conductor, multiply the length times the width of the end face of the conductor (side is expressed in mils). For example, assume that one side of the rectangular cross-sectional area is 6 mils and the other side is 3 mils. Multiply 6 mils × 3 mils, which equals 18 square mils. Here is another example. Assume that a conductor is 3/8 inch thick and 4 inches wide. The 3/8 inch can be expressed in decimal form as 0.375 inch. Since 1 mil equals 0.001 inch, the thickness of the conductor will be 0.001 × 0.375, or 375 mils. Since the width is 4 inches and there are 1,000 mils per inch, the width will be 4 × 1,000, or 4,000 mils. To determine the cross-sectional area, multiply the length by the width; or 375 mils × 4,000 mils. The area will be 1,500,000 square mils.

Q4. Define a square mil as it relates to a square conductor.

CIRCULAR MIL

The circular mil is the standard unit of measurement of a round wire cross-sectional area (view C of figure 1-1). This unit of measurement is found in American and English wire tables. The diameter of a round conductor (wire) used to conduct electricity may be only a fraction of an inch. Therefore, it is convenient to express this diameter in mils to avoid using decimals. For example, the diameter of a wire is expressed as 25 mils instead of 0.025 inch. A circular mil is the area of a circle having a diameter of 1 mil, as shown in view B of figure 1-2. The area in circular mils of a round conductor is obtained by squaring the diameter, measured in mils. Thus, a wire having a diameter of 25 mils has an area of 25^2, or 625 circular mils. To determine the number of square mils in the same conductor, apply the conventional formula for determining the area of a circle ($A = \pi r^2$). In this formula, A (area) is the unknown and is equal to the cross-sectional area in square mils, π is the constant 3.14, and r is the radius of the circle, or half the diameter (D). Through substitution, A = 3.14, and $(12.5)^2$; therefore, $3.14 \times 156.25 = 490.625$

square mils. The cross-sectional area of the wire has 625 circular mils but only 490.625 square mils. Therefore, a circular mil represents a smaller unit of area than the square mil.

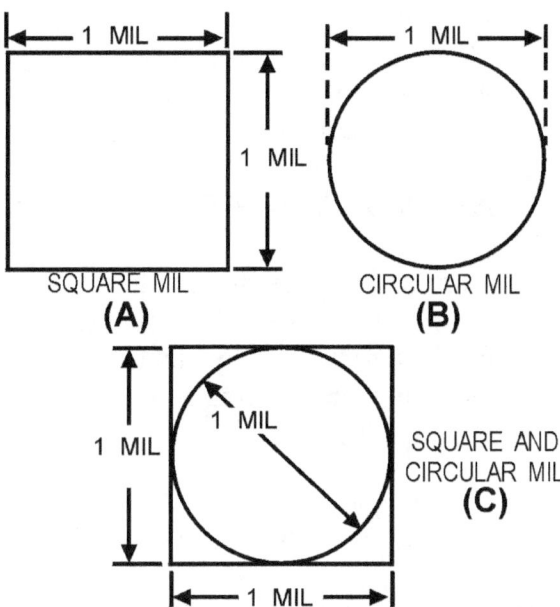

Figure 1-2.—A comparison of circular and square mils.

If a wire has a cross-sectional diameter of 1 mil, by definition, the circular mil area (CMA) is $A = D^2$, or $A = 1^2$, or $A = 1$ circular mil. To determine the square mil area of the same wire, apply the formula $A = \pi r^2$; therefore, $A = 3.14 \times (.5)^2$ (.5 representing half the diameter). When $A = 3.14 \times .25$, $A = .7854$ square mil. From this, it can be concluded that 1 circular mil is equal to. 7854 square mil. This becomes important when square (view A of figure 1-2) and round (view B) conductors are compared as in view C of figure 1-2.

When the square mil area is given, divide the area by 0.7854 to determine the circular mil area, or CMA. When the CMA is given, multiply the area by 0.7854 to determine the square mil area. For example,

Problem: A 12-gauge wire has a diameter of 80.81 mils. What is (1) its area in circular mils and (2) its area in square mils?

Solution:

(1) $A = D^2 = 80.81^2 = 6,530$ circular mils
(2) $A = 0.7854 \times 6,530 = 5,128.7$ square mils

Problem: A rectangular conductor is 1.5 inches wide and 0.25 inch thick. What is (1) its area in square mils and (2) in circular mils? What size of round conductor is necessary to carry the same current as the rectangular bar?

Solution:

(1) 1.5 inches = 1.5 inches × 1,000
mils per inch = 1,500 mils

0.25 inch = 0.25 inch × 1,000 mils
per inch = 250 mils

A = 1,500 × 250 = 375,000 square mils

(2) To carry the same current, the
cross-sectional area of the round
conductor must be equal. There are
more circular mils than square mils in
this area. Therefore:

$$A = \frac{375,000}{0.7854} = 477,000 \text{ circular mils}$$

A wire in its usual form is a single slender rod or filament of drawn metal. In large sizes, wire becomes difficult to handle. To increase its flexibility, it is stranded. Strands are usually single wires twisted together in sufficient numbers to make up the necessary cross-sectional area of the cable. The total area of stranded wire in circular mils is determined by multiplying the area in circular mils of one strand by the number of strands in the cable.

Q5. *Define a circular mil.*

Q6. *What is the circular mil area of a 19-strand conductor if each strand is 0.004 inch?*

CIRCULAR-MIL-FOOT

A circular-mil-foot (figure 1-3) is a unit of volume. It is a unit conductor 1 foot in length and has a cross-sectional area of 1 circular mil. Because it is a unit conductor, the circular-mil-foot is useful in making comparisons between wires consisting of different metals. For example, a basis of comparison of the RESISTIVITY (to be discussed shortly) of various substances may be made by determining the resistance of a circular-mil-foot of each of the substances.

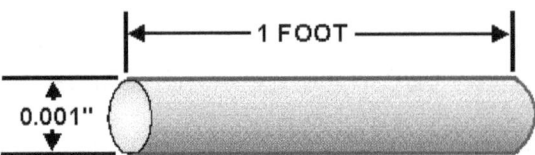

Figure 1-3.—Circular-mil-foot.

In working with square or rectangular conductors, such as ammeter shunts and bus bars, you may sometimes find it more convenient to use a different unit volume. A bus bar is a heavy copper strap or bar used to connect several circuits together. Bus bars are used when a large current capacity is required. Unit volume may be measured as the centimeter cube. Specific resistance, therefore, becomes the resistance offered by a cube-shaped conductor 1centimeter in length and 1 square centimeter in cross-sectional area. The unit of volume to be used is given in tables of specific resistances.

SPECIFIC RESISTANCE OR RESISTIVITY

Specific resistance, or resistivity, is the resistance in ohms offered by a unit volume (the circular-mil-foot or the centimeter cube) of a substance to the flow of electric current. Resistivity is the reciprocal of conductivity. A substance that has a high resistivity will have a low conductivity, and vice versa. Thus, the specific resistance of a substance is the resistance of a unit volume of that substance.

Many tables of specific resistance are based on the resistance in ohms of a volume of a substance 1 foot in length and 1 circular mil in cross-sectional area. The temperature at which the resistance measurement is made is also specified. If you know the kind of metal a conductor is made of, you can obtain the specific resistance of the metal from a table. The specific resistances of some common substances are given in table 1-1.

Table 1-1.—Specific Resistances of Common Substances

Substance	Specific resistance at 20°C.	
	Centimeter cube (microhoms)	Circular-mil-foot (ohms)
Silver	1.629	9.8
Copper (drawn)	1.724	10.37
Gold	2.44	14.7
Aluminum	2.828	17.02
Carbon (amorphous)	3.8 to 4.1	....
Tungsten	5.51	33.2
Brass	7.0	42.1
Steel (soft)	15.9	95.8
Nichrome	109.0	660.0

The resistance of a conductor of a uniform cross section varies directly as the product of the length and the specific resistance of the conductor, and inversely as the cross-sectional area of the conductor. Therefore, you can calculate the resistance of a conductor if you know the length, cross-sectional area, and specific resistance of the substance. Expressed as an equation, the "R" (resistance in ohms) of a conductor is

$$R = \rho \frac{L}{A}$$

Where:

ρ = (Greek rho) the specific
resistance in ohms per
circular-mil-foot (refer to table 1-1)

L = length in feet

A = the cross-sectional area in
circular mils

Problem:

What is the resistance of 1,000 feet of copper wire having a cross-sectional area of 10,400 circular mils (No. 10 wire) at a temperature of 20° C?

Solution:

The specific resistance of copper (table 1-1) is 10.37 ohms. Substituting the known values in the preceding equation, the resistance, R, is determined as

Given: ρ = 10.37 ohms

L = 1,000 ft

A = 10,400 circular mils

Solution:
$$R = \rho \frac{L}{A} = 10.37 \times \frac{1,000}{10,400}$$

= 1 ohm (approximately)

If R, ρ, and A are known, the length (L) can be determined by a simple mathematical transposition. This has many valuable applications. For example, when locating a ground in a telephone line, you will use special test equipment. This equipment operates on the principle that the resistance of a line varies directly with its length. Thus, the distance between the test point and a fault can be computed accurately.

Q7.   *Define specific resistance.*

Q8.   *List the three factors used to calculate resistance of a particular conductor in ohms.*

WIRE SIZES

The most common method for measuring wire size in the Navy is by using the American Wire Gauge (AWG). An exception is aircraft wiring, which varies slightly in size and flexibility from AWG standards. For information concerning aircraft wire sizes, refer to the proper publications for specific aircraft. Only AWG wire sizes are used in the following discussion.

Wire is manufactured in sizes numbered according to the AWG tables. The various wires (solid or stranded) and the material they are made from (copper, aluminum, and so forth) are published by the National Bureau of Standards. An AWG table for copper wire is shown at table 1-2. The wire diameters become smaller as the gauge numbers become larger. Numbers are rounded off for convenience but are accurate for practical application. The largest wire size shown in the table is 0000 (read "4 naught"), and the smallest is number 40. Larger and smaller sizes are manufactured, but are not commonly used by the Navy. AWG tables show the diameter in mils, circular mil area, and area in square inches of AWG wire sizes. They also show the resistance (ohms) per thousand feet and per mile of wire sizes at specific temperatures. The last column shows the weight of the wire per thousand feet. An example of the use of table 1-2 is as follows.

Table 1-2.—Standard Solid Copper (American Wire Gauge)

Gage number	Diameter (mils)	Cross Section		Ohms per 1,000 ft		Ohms per mile 25°C. = 77°F.	Pounds per 1,000 ft.
		Circular mils	Square inches	25°C. = 77°F.	65°C. = 149°F.		
0000	460.0	212,000.0	0.166	0.0500	0.0577	0.264	641.0
000	410.0	168,000.0	.132	.0630	.0727	.333	508.0
00	365.0	133,000.0	.105	.0795	.0917	.420	403.0
0	325.0	106,000.0	.0829	.100	.116	.528	319.0
1	289.0	83,700.0	.0657	.126	.146	.665	253.0
2	258.0	66,400.0	.0521	.159	.184	.839	201.0
3	229.0	52,600.0	.0413	.201	.232	1.061	159.0
4	204.0	41,700.0	.0328	.253	.292	1.335	126.0
5	182.0	33,100.0	.0260	.319	.369	1.685	100.0
6	162.0	26,300.0	.0206	.403	.465	2.13	79.5
7	144.0	20,800.0	.0164	.508	.586	2.68	63.0
8	128.0	16,500.0	.0130	.641	.739	3.38	50.0
9	114.0	13,100.0	.0103	.808	.932	4.27	39.6
10	102.0	10,400.0	.00815	1.02	1.18	5.38	31.4
11	91.0	8,230.0	.00647	1.28	1.48	6.75	24.9
12	81.0	6,530.0	.00513	1.62	1.87	8.55	19.8
13	72.0	5,180.0	.00407	2.04	2.36	10.77	15.7
14	64.0	4,110.0	.00323	2.58	2.97	13.62	12.4
15	57.0	3,260.0	.00256	3.25	3.75	17.16	9.86
16	51.0	2,580.0	.00203	4.09	4.73	21.6	7.82
17	45.0	2,050.0	.00161	5.16	5.96	27.2	6.20
18	40.0	1,620.0	.00128	6.51	7.51	34.4	4.92
19	36.0	1,290.0	.00101	8.21	9.48	43.3	3.90
20	32.0	1,020.0	.000802	10.4	11.9	54.9	3.09
21	28.5	810.0	.000636	13.1	15.1	69.1	2.45
22	25.3	642.0	.000505	16.5	19.0	87.1	1.94
23	22.6	509.0	.000400	20.8	24.0	109.8	1.54
24	20.1	404.0	.000317	26.2	30.2	138.3	1.22
25	17.9	320.0	.000252	33.0	38.1	174.1	0.970
26	15.9	254.0	.000200	41.6	48.0	220.0	0.769
27	14.2	202.0	.000158	52.5	60.6	277.0	0.610
28	12.6	160.0	.000126	66.2	76.4	350.0	0.484
29	11.3	127.0	.0000995	83.4	96	440.0	0.384
30	10.0	101.0	.0000789	105.0	121.0	554.0	0.304
31	8.9	79.7	.0000626	133.0	153.0	702.0	0.241
32	8.0	63.2	.0000496	167.0	193.0	882.0	0.191
33	7.1	50.1	.0000394	211.0	243.0	1,114.0	0.152
34	6.3	39.8	.0000312	266.0	307.0	1,404.0	0.120
35	5.6	31.5	.0000248	335.0	387.0	1,769.0	0.0954
36	5.0	25.0	.0000196	423.0	488.0	2,230.0	0.0757
37	4.5	19.8	.0000156	533.0	616.0	2,810.0	0.0600
38	4.0	15.7	.0000123	673.0	776.0	3,550.0	0.0476
39	3.5	12.5	.0000098	848.0	979.0	4,480.0	0.0377
40	3.1	9.9	.0000078	1,070.0	1,230.0	5,650.0	0.0299

Problem: You are required to run 2,000 feet of AWG 20 solid copper wire for a new piece of equipment. The temperature where the wire is to be run is 25° C (77° F). How much resistance will the wire offer to current flow?

Solution: Under the gauge number column, find size AWG 20. Now read across the columns until you reach the "ohms per 1,000 feet for 25° C (77° F)" column. You will find that the wire will offer 10.4 ohms of resistance to current flow. Since we are using 2,000 feet of wire, multiply by 2.

10.4 ohms × 2 = 20.8 ohms

An American Standard Wire Gauge (figure 1-4) is used to measure wires ranging in size from number 0 to number 36. To use this gauge, insert the wire to be measured into the smallest slot that will just accommodate the bare wire. The gauge number on that slot indicates the wire size. The front part of the slot has parallel sides, and this is where the wire measurement is taken. It should not be confused with the larger semicircular opening at the rear of the slot. The rear opening simply permits the free movement of the wire all the way through the slot.

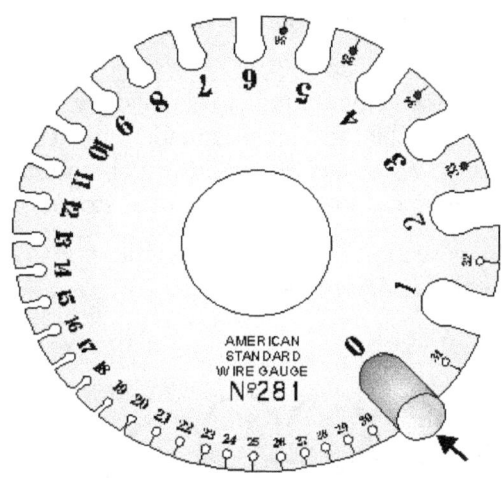

AMERICAN
STANDARD
WIRE GAUGE
N°281

Figure 1-4.—Wire gauge.

Q9.   Using table 1-2, determine the resistance of 1,500 feet of AWG 20 wire at 25° C.

Q10.   When using an American Standard Wire Gauge to determine the size of a wire, where should you place the wire in the gauge to get the correct measurement?

STRANDED WIRES AND CABLES

A <u>wire</u> is a single slender rod or filament of drawn metal. This definition restricts the term to what would ordinarily be understood as "solid wire." The word "slender" is used because the length of a wire is usually large when compared to its diameter. If a wire is covered with insulation, it is an insulated wire. Although the term "wire" properly refers to the metal, it also includes the insulation.

A <u>conductor</u> is a wire suitable for carrying an electric current.

A <u>stranded conductor</u> is a conductor composed of a group of wires or of any combination of groups of wires. The wires in a stranded conductor are usually twisted together and not insulated from each other.

A <u>cable</u> is either a stranded conductor (single-conductor cable) or a combination of conductors insulated from one another (multiple-conductor cable). The term "cable" is a general one and usually applies only to the larger sizes of conductors. A small cable is more often called a stranded wire or cord (such as that used for an iron or a lamp cord). Cables may be bare or insulated. Insulated cables may be sheathed (covered) with lead, or protective armor. Figure 1-5 shows different types of wire and cable used in the Navy.

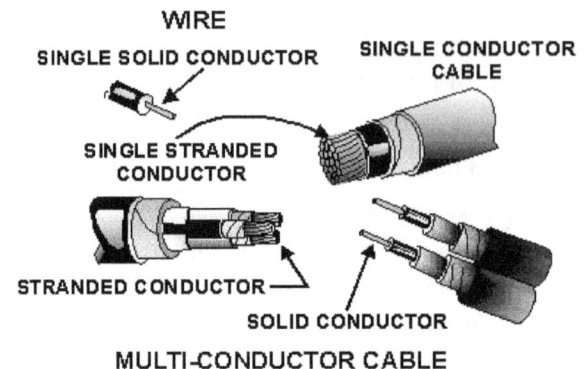

Figure 1-5.—Conductors.

Conductors are stranded mainly to increase their flexibility. The wire strands in cables are arranged in the following order:

The first layer of strands around the center conductor is made up of six conductors. The second layer is made up of 12 additional conductors. The third layer is made up of 18 additional conductors, and so on. Thus, standard cables are composed of 7, 19, and 37 strands, in continuing fixed increments. The overall flexibility can be increased by further stranding of the individual strands.

Figure 1-6 shows a typical cross section of a 37-strand cable. It also shows how the total circular-mil cross-sectional area of a stranded cable is determined.

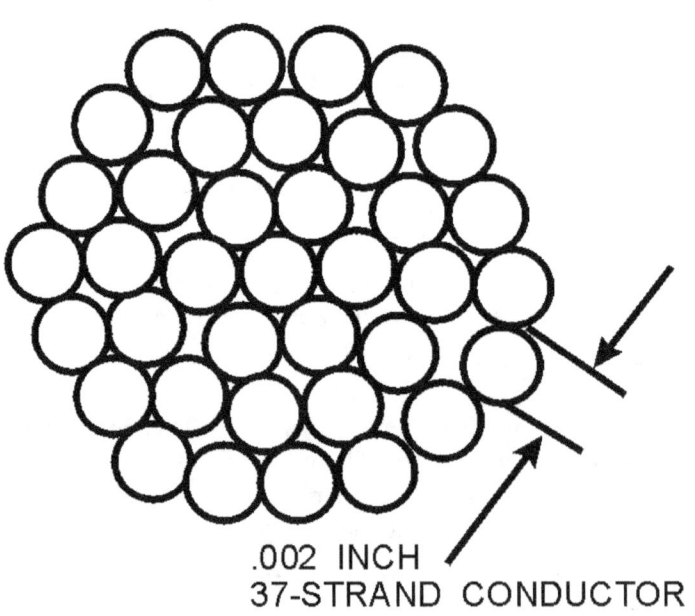

DIAMETER OF EACH STRAND = .002 INCH
DIAMETER OF EACH STRAND = 2 MILS
CIRCULAR MIL AREA OF EACH STRAND = D2 = 4 CM
TOTAL CM AREA OF CONDUCTOR = 4X37 = 148 CM

Figure 1-6.—Stranded conductor.

SELECTION OF WIRE SIZE

Several factors must be considered in selecting the size of wire to be used for transmitting and distributing electric power. These factors will be discussed throughout this section. Military specifications cover the installation of wiring in aircraft, ships, and electrical/electronic equipment. These specifications describe the technical requirements for material purchased from manufacturers by the Department of Defense. An important reason for having these specifications is to ensure uniformity of sizes to reduce the danger of fires caused by the improper selection of wire sizes. Wires can carry only a limited amount of current safely. If the current flowing through a wire exceeds the current-carrying capacity of the wire, excess heat is generated. This heat may be great enough to burn off the insulation around the wire and start a fire.

FACTORS GOVERNING THE CURRENT RATING

The current rating of a cable or wire indicates the current capacity that the wire or cable can safely carry continuously. If this limit, or current rating, is exceeded for a length of time, the heat generated may burn the insulation. The current rating of a wire is used to determine what size is needed for a given load, or current drain.

The factors that determine the current rating of a wire are the conductor size, the location of the wire in a circuit, the type of insulation, and the safe current rating. Another factor that will be discussed later in this chapter is the material the wire is made of. As you have already seen, these factors also affect the resistance in ohms of a wire-carrying current.

CONDUCTOR SIZE

An increase in the diameter, or cross section, of a wire conductor decreases its resistance and increases its capacity to carry current. An increase in the specific resistance of a conductor increases its resistance and decreases its capacity to carry current.

WIRE LOCATION

The location of a wire in a circuit determines the temperature under which it operates. A wire may be located in a conduit or laced with other wires in a cable. Because it is confined, the wire operates at a higher temperature than if it were open to the free air. The higher the temperature under which a wire is operating, the greater will be its resistance. Its capacity to carry current is also lowered. Note that, in each case, the resistance of a wire determines its current-carrying capacity. The greater the resistance, the more power it dissipates in the form of heat energy.

Conductors may also be installed in locations where the ambient (surrounding) temperature is relatively high. When this is the case, the heat generated by external sources is an important part of the total conductor heating. This heating factor will be explained further when we discuss temperature coefficient. We must understand how external heating influences how much current a conductor can carry. Each case has its own specific limitations. The maximum allowable operating temperature of insulated conductors is specified in tables. It varies with the type of conductor insulation being used.

INSULATION

The insulation of a wire does not affect the resistance of the wire. Resistance does, however, determine how much heat is needed to burn the insulation. As current flows through an insulated conductor, the limit of current that the conductor can withstand depends on how hot the conductor can get before it burns the insulation. Different types of insulation will burn at different temperatures. Therefore, the type of insulation used is the third factor that determines the current rating of a conductor. For instance, rubber insulation will begin deteriorating at relatively low temperatures, whereas varnished cloth insulation retains its insulating properties at higher temperatures. Other types of insulation are fluorinated ethylene propylene (FEP), silicone rubber, or extruded polytetrafluoroethylene. They are effective at still higher temperatures.

SAFE CURRENT RATINGS

The National Board of Fire Underwriters prepares tables showing the safe current ratings for sizes and types of conductors covered with various types of insulation. The allowable current-carrying capacities of single copper conductors in free air at a maximum room temperature of 30° C (86° F) are given in table 1-3. At ambient temperatures greater than 30° C, these conductors would have less current-carrying capacity.

Table 1-3.—Temperature Ratings and Current-Carrying Capacities (in Amperes) of Some Single Copper Conductors at Ambient Temperatures of 30°C

Size	Moisture Resistant Rubber or Thermoplastic	Varnished Cambric or Heat Resistant Thermoplastic	Silicone Rubber or Fluorinated Ethylene Propylene (FEP)	Polytetra-Fluoroethylene
0000	300	385	510	850
000	260	330	430	725
00	225	285	370	605
0	195	245	325	545
1	165	210	280	450
2	140	180	240	390
3	120	155	210	335
4	105	135	180	285
6	80	100	135	210
8	55	70	100	115
10	40	55	75	110
12	25	40	55	80
14	20	30	45	60

Q11. *List the four factors you should use to select wire for a specified current rating.*

Q12. *What are three types of nonmetallic insulating materials that can be used in a high-temperature environments?*

Q13. *State why it is important for you to consider the ambient (surrounding) temperature of a conductor when selecting wire size.*

COPPER-VERSUS-ALUMINUM CONDUCTORS

Although silver is the best conductor, its cost limits its use to special circuits. Silver is used where a substance with high conductivity or low resistivity is needed.

The two most commonly used conductors are copper and aluminum. Each has positive and negative characteristics that affect its use under varying circumstances. A comparison of some of the characteristics of copper and aluminum is given in table 1-4.

Table 1-4.—Comparative Characteristics of Copper and Aluminum

CHARACTERISTICS	COPPER	ALUMINUM
Tensile strength (lb/in2).	55,000	25,000
Tensile strength for same conductivity (lb).	55,000	40,000
Weight for same conductivity (lb).	100	48
Cross section for same conductivity (C.M.).	100	160
Specific resistance (Ω/mil ft).		

Copper has a higher conductivity than aluminum. It is more ductile (can be drawn out). Copper has relatively high tensile strength (the greatest stress a substance can bear along its length without tearing apart). It can also be easily soldered. However, copper is more expensive and heavier than aluminum.

Although aluminum has only about 60 percent of the conductivity of copper, its lightness makes long spans possible. Its relatively large diameter for a given conductivity reduces corona. Corona is the discharge of electricity from the wire when it has a high potential. The discharge is greater when smaller diameter wire is used than when larger diameter wire is used. However, the relatively large size of aluminum for a given conductance does not permit the economical use of an insulation covering.

Q14. State two advantages of using aluminum wire for carrying electricity over long distances.

Q15. State four advantages of copper over aluminum as a conductor.

TEMPERATURE COEFFICIENT

The resistance of pure metals, such as silver, copper, and aluminum, increases as the temperature increases. However, the resistance of some alloys, such as constantan and manganin, changes very little as the temperature changes. Measuring instruments use these alloys because the resistance of the circuits must remain constant to get accurate measurements.

In table 1-1, the resistance of a circular-mil-foot of wire (the specific resistance) is given at a specific temperature, 20° C in this case. It is necessary to establish a standard temperature. As we stated earlier, the resistance of pure metals increases with an increase in temperature. Therefore, a true basis of comparison cannot be made unless the resistances of all the substances being compared are measured at the same temperature. The amount of increase in the resistance of a 1-ohm sample of the conductor per degree rise in temperature above 0° C is called the temperature coefficient of resistance. For copper, the value is approximately 0.00427 ohm.

A length of copper wire having a resistance of 50 ohms at an initial temperature of 0° C will have an increase in resistance of 50×0.00427, or 0.214 ohms. This applies to the entire length of wire and for each degree of temperature rise above 0° C. A 20° C increase in resistance is approximately 20×0.214, or 4.28 ohms. The total resistance at 20° C is 50 + 4.28, or 54.28 ohms.

Q16. Define the temperature coefficient of resistance.

Q17. What happens to the resistance of copper when it is heated?

CONDUCTOR INSULATION

To be useful and safe, electric current must be forced to flow only where it is needed. It must be "channeled" from the power source to a useful load. In general, current-carrying conductors must not be allowed to come in contact with one another, their supporting hardware, or personnel working near them.

To accomplish this, conductors are coated or wrapped with various materials. These materials have such a high resistance that they are, for all practical purposes, nonconductors. Nonconductors are generally referred to as "insulators" or "insulating material."

Only the necessary minimum amount of insulation is applied to any particular type of conductor designed to do a particular job. This is done because of several factors. The expense, stiffening effect, and a variety of physical and electrical conditions under which the conductors are operated must be taken into account. Therefore, there are a variety of insulated conductors available to meet the requirements of any job.

Two fundamental properties of insulating materials (that is, rubber, glass, asbestos, or plastic) are insulation resistance and dielectric strength. These are two entirely different and distinct properties.

INSULATION RESISTANCE

Insulation resistance is the resistance to current leakage through the insulation materials. Insulation resistance can be measured with a megger without damaging the insulation. Information so obtained is a useful guide in appraising the general condition of insulation. Clean, dry insulation having cracks or other faults may show a high value of insulation resistance but would not be suitable for use.

DIELECTRIC STRENGTH

Dielectric strength is the ability of an insulator to withstand potential difference. It is usually expressed in terms of the voltage at which the insulation fails because of the electrostatic stress. Maximum dielectric strength values can be measured only by raising the voltage of a TEST SAMPLE until the insulation breaks down.

Q18.   Compare the resistance of a conductor to that of an insulator.

Q19.   State two fundamental properties of insulating materials.

Q20.   Define insulation resistance.

Q21.   Define dielectric strength.

Q22.   How is the dielectric strength of an insulator determined?

TYPES OF INSULATION

The insulating materials discussed in the next paragraphs are commonly used in Navy electrical and electronic equipment.

Rubber

One of the most common types of insulation is rubber. The voltage that may be applied to a rubber-covered conductor is dependent on the thickness and the quality of the rubber covering. Other factors being equal, the thicker the insulation, the higher may be the applied voltage. Rubber insulation is normally used for low- or medium-range voltage. Figure 1-7 shows two types of rubber-covered wire. One is a two-conductor cable in which each stranded conductor is covered with rubber insulation; the other is a single, solid conductor. In each case, the rubber serves the same purpose: to confine the current to its conductor.

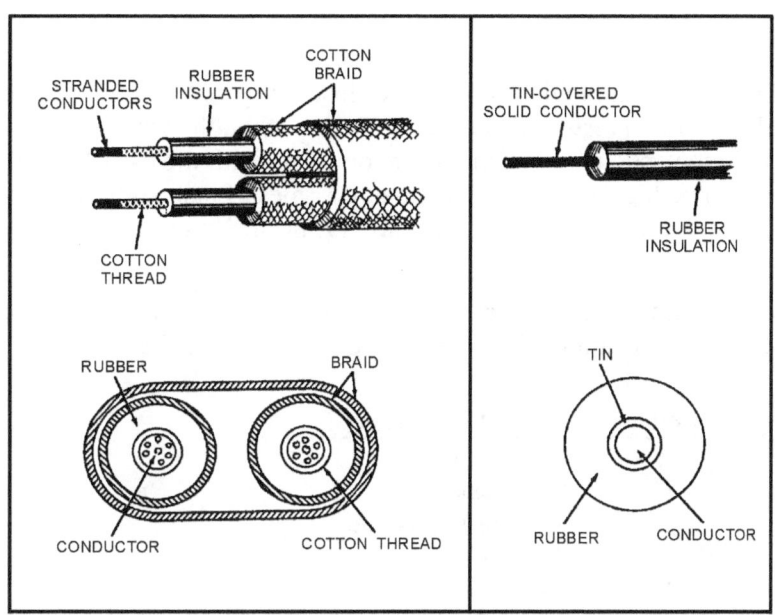

Figure 1-7.—Rubber insulation.

Referring to the enlarged cross-sectional view in figure 1-7, note that a thin coating of tin separates the copper conductor from the rubber insulation. If the thin coating of tin were not used, a chemical action would take place and the rubber would become soft and gummy where it makes contact with the copper. When small, solid, or stranded conductors are used, a winding of cotton threads is applied between the conductors and the rubber insulation.

CODE-GRADED RUBBER.—Code-graded rubber is the standard that the National Electrical Code (NEC) has adopted as the minimum requirements for rubber insulation as specified by Underwriters' Laboratories. In this code system, the letter R indicates the use of a rubber insulator. Type R signifies that the wire is rubber coated.

The NEC codes Type RH and Type RHH signify a rubber heat-resistant compound. Type RW signifies a rubber moisture-resistant compound. A Type RHW signifies a rubber heat- and moisture-resistant compound. Type RHW is approved for use in wet or dry locations at a maximum conductor temperature of 75° C. Neoprene, a low-voltage compound, is the one exception to Type RHW. Although not a rubber compound, neoprene meets the requirements of Underwriters' Laboratories and was designated Type RHW.

LATEX RUBBER.—Latex rubber is a high-grade compound consisting of 90 percent unmilled grainless rubber. There are two designations for this type of insulation: Type RUH and Type RUW. Type RUH (rubber unmilled heat-resistant) is used in dry locations when the conductor temperature does not exceed 75° C. Type RUW (rubber unmilled moisture-resistant) is used in wet locations when the conductor does not exceed 60° C.

SILICONE.—Silicone is a rubber compound that does not carry the "R" designator for many of its applications. An example of this is Type SA (silicone-asbestos). In Type SA, the insulator around the conductor is silicone rubber, but the outer covering must consist of heavy glass, asbestos-glass, or asbestos braiding treated with a heat, flame, and moisture-resistant compound.

Q23.   What is the purpose of coating a copper conductor with tin when rubber insulation is used?

Plastics

Plastic is one of the more commonly used types of insulating material for electrical conductors. It has good insulating, flexibility, and moisture-resistant qualities. Although there are many types of plastic insulating materials, thermoplastic is one of the most common. With the use of thermoplastic, the conductor temperature can be higher than with some other types of insulating materials without damage to the insulating quality of the material. Plastic insulation is normally used for low- or medium-range voltage.

The designators used with thermoplastics are much like those used with rubber insulators. The following letters are used when dealing with NEC type designators for thermoplastics:

Letter	Designates
T	Thermoplastic
H	Heat-resistant
W	Moisture-resistant
A	Asbestos
N	Outer nylon jacket
M	Oil-resistant

For example, a NEC designator of Type THWN would indicate thermoplastic heat- and moisture-resistant with an outer nylon jacket.

Varnished Cambric

Varnished cambric insulation can withstand much higher temperatures than rubber insulation. Varnished cambric is cotton cloth that has been coated with an insulating varnish. Figure 1-8 shows a cable covered with varnished cambric insulation. The varnished cambric is in tape form and is wound around the conductor in layers. An oily compound is applied between each layer of the tape to prevent water from seeping through the insulation. It also acts as a lubricant between the layers of tape, so they will slide over each other when the cable is bent.

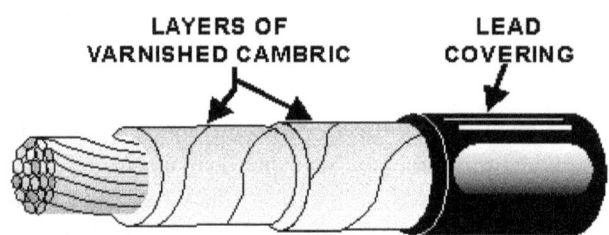

Figure 1-8.—Varnished cambric insulation.

Cambric insulation is used on extremely high-voltage conductors used in substations and powerhouses. It is also used in other locations subjected to high temperatures. In addition, it is used on the coils and leads of high-voltage generators. Transformer leads also use this insulation because it is unaffected by oils or grease and has high dielectric strength. Varnished cambric and paper insulation for cables are the two types of insulating materials most widely used at voltages above 15,000 volts. Such cable is always lead covered to keep out moisture.

Extruded Polytetrafluoroethylene

Extruded polytetrafluoroethylene is a high-temperature insulation used extensively in aircraft and equipment installations. It will not burn, but will vaporize when subjected to intense heat. Conductors for high temperatures use a nickel coating rather than tin or silver to prevent oxidation. Nickel-coated wire is more difficult to solder, but makes satisfactory connections with proper soldering techniques.

WARNING

Avoid breathing the vapors from extruded polytetrafluoroethylene insulation when it is heated. Symptoms of overexposure are dizziness or headaches. These symptoms disappear upon exposure to fresh air.

Q24.   What safety precaution should you take when working with extruded polytetrafluoroethylene insulated wiring?

Fluorinated Ethylene Propylene (FEP)

FEP has properties similar to extruded polytetrafluoroethylene, but will melt at soldering temperatures. It is rated at 200° C and is, therefore, considered a high-temperature insulation. There are no known toxic vapors from FEP. Common-sense practice, however, requires that you provide adequate ventilation during any soldering operation.

Asbestos

Asbestos insulation was used extensively in the past for high-temperature insulation. Today, it is seldom used by the Navy. Many naval ships and aircraft, however, still contain asbestos-insulated wiring. Aboard ship, this is particularly true in galley and laundry equipment. The reason for discontinuing the use of asbestos as an insulator is that breathing asbestos fibers can produce severe lung damage. It can render you disabled or cause fatal fibrosis of the lungs. Asbestos is also a factor in the development of cancer in the gastrointestinal tract. Safety precautions concerning asbestos will be covered in more detail at the end of chapter 3.

WARNING

Avoid inhalation of asbestos fibers. Asbestos fibers have been found to cause severe lung damage (asbestosis) and cancer of the gastrointestinal tract. Follow Navy safety precautions when working with all asbestos products.

One type of asbestos-covered wire is shown in figure 1-9. It consists of stranded copper conductors covered with felted asbestos. The wire is, in turn, covered with asbestos braid. This type of wire is used in motion-picture projectors, arc lamps, spotlights, heating element leads, and so forth.

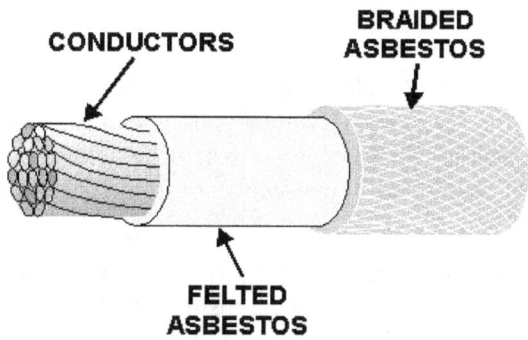

Figure 1-9.—Asbestos Insulation.

Another type of asbestos-covered cable is shown in figure 1-10. It is combination of asbestos and varnished cambric. This type of insulation serves as leads for motors and transformers that sometimes must operate in hot, damp locations. The varnished cambric covers the inner layer of felted asbestos. This prevents moisture from reaching the innermost layer of asbestos. Asbestos loses its insulating properties when it becomes wet. It will, in fact, become a conductor. Varnished cambric prevents this from happening because it resists moisture. Although this insulation will withstand some moisture, it should not be used on conductors that may at times be partially immersed in water. Under those circumstances, the insulation must be protected with an outer lead sheath.

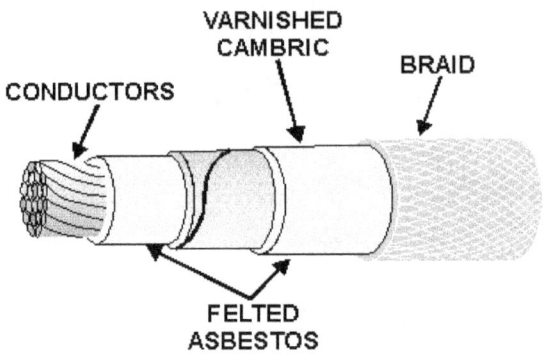

Figure 1-10.—Asbestos and varnished cambric insulation.

The NEC has designators for eight types of asbestos wire. The designators and a description of each are listed below.

Type A	Nonimpregnated asbestos without an asbestos braid
Type AA	Nonimpregnated asbestos with an outer asbestos braid or glass
Type AI	Impregnated asbestos without an asbestos braid
Type AIA	Impregnated asbestos with an outer asbestos braid or glass
Type AVA	Asbestos, varnish-cambric insulation with an outer asbestos braid or glass
Type AVL	Asbestos, varnish-cambric insulation with an outer asbestos braid covered with a lead sheath
Type AVB	Asbestos, varnish-cambric insulation with an outer flame-retardant cotton braid
Type SA	Silicone rubber insulated with outer heavy glass, asbestos-glass, or asbestos braid

Q25. State the reasons that the Navy is getting away from the use of asbestos insulation.

Q26. State what happens to the insulating characteristics of asbestos when it gets wet.

Paper

Paper has little insulation value alone. However, when impregnated with a high grade of mineral oil, it serves as a satisfactory insulation for extremely high-voltage cables. The oil has a high dielectric strength, and tends to prevent breakdown of the paper insulation. The paper must be thoroughly saturated with the oil. The thin paper tape is wrapped in many layers around the conductors, and then soaked with oil.

The three-conductor cable shown in figure 1-11 consists of paper insulation on each conductor. It has a spirally wrapped nonmagnetic metallic tape over the insulation. The space between conductors is filled with a suitable spacer to round out the cable. Another nonmagnetic metal tape is used to secure the entire cable. Over this, a lead sheath is applied. This type of cable is used on voltages from 10,000 volts to 35,000 volts.

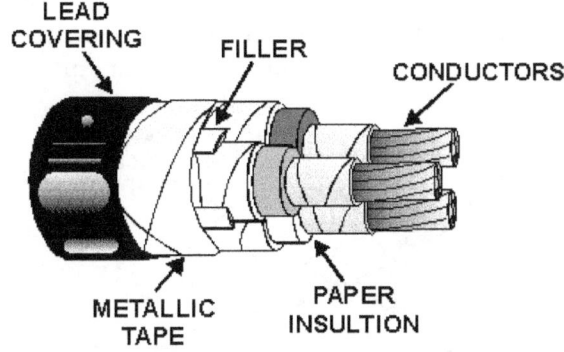

Figure 1-11.—Paper-insulated power cables.

Q27. What are the most common insulators used for extremely high voltages?

Silk and Cotton

In certain types of circuits (for example, communications circuits), a large number of conductors are needed, perhaps as many as several hundred. Figure 1-12 shows a cable containing many conductors. Each is insulated from the others by silk and cotton thread. Because the insulation in this type of cable is not subjected to high voltage, the use of thin layers of silk and cotton is satisfactory.

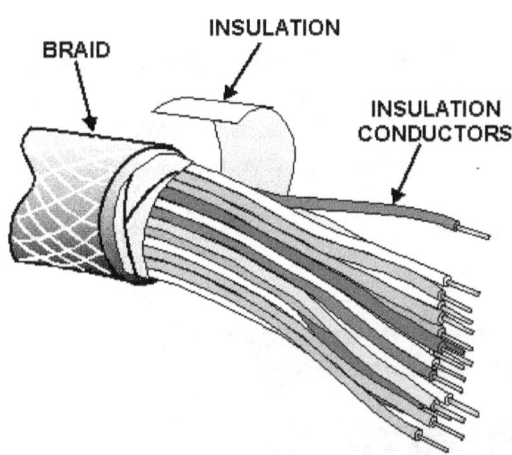

Figure 1-12.—Silk and cotton Insulation.

Silk and cotton insulation keeps the size of the cable small enough to be handled easily. The silk and cotton threads are wrapped around the individual conductors in reverse directions. The covering is then impregnated with a special wax compound.

Enamel

The wire used on the coils of meters, relays, small transformers, motor windings, and so forth, is called magnet wire. This wire is insulated with an enamel coating. The enamel is a synthetic compound of cellulose acetate (wood pulp and magnesium). In the manufacturing process, the bare wire is passed through a solution of hot enamel and then cooled. This process is repeated until the wire acquires from 6 to 10 coatings. Thickness for thickness, enamel has higher dielectric strength than rubber. It is not practical for large wires because of the expense and because the insulation is readily fractured when large wires are bent.

Figure 1-13 shows an enamel-coated wire. Enamel is the thinnest insulating coating that can be applied to wires. Hence, enamel-insulated magnet wire makes smaller coils. Enameled wire is sometimes covered with one or more layers of cotton to protect the enamel from nicks, cuts, or abrasions.

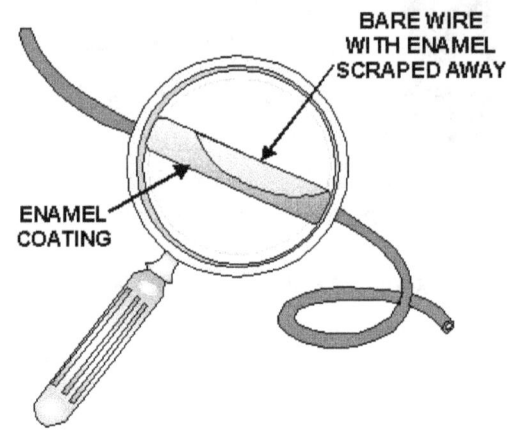

Figure 1-13.—Enamel Insulation.

Q28.   *What is the common name for enamel-insulated wire?*

Mineral Insulated

Mineral-insulated (MI) cable was developed to meet the needs of a noncombustible, high heat-resistant, and water-resistant cable. MI cable has from one to seven electrical conductors. These conductors are insulated in a highly compressed mineral, normally magnesium oxide, and sealed in a liquidtight, gastight metallic tube, normally made of seamless copper (figure 1-14).

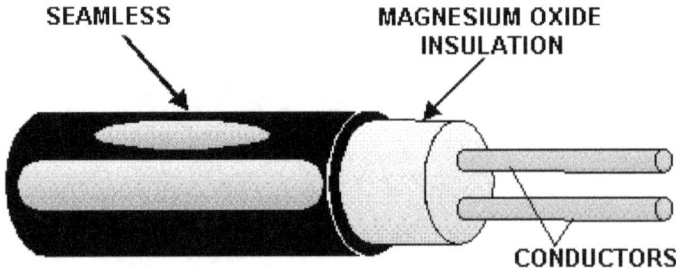

Figure 1-14.—Two-conductor mineral-insulated (MI) cable.

CONDUCTOR PROTECTION

Wires and cables are generally subject to abuse. The type and amount of abuse depends on how and where they are installed and the manner in which they are used. Cables buried directly in the ground must resist moisture, chemical action, and abrasion. Wires installed in buildings must be protected against mechanical injury and overloading. Wires strung on crossarms on poles must be kept far enough apart so that the wires do not touch. Snow, ice, and strong winds make it necessary to use conductors having high tensile strength and substantial frame structures.

Generally, except for overhead transmission lines, wires or cables are protected by some form of covering. The covering may be some type of insulator like rubber or plastic. Over this, additional layers of fibrous braid or tape may be used and then covered with a finish or saturated with a protective coating. If the wire or cable is installed where it is likely to receive rough treatment, a metallic coat should be added.

The materials used to make up the protection for a wire or cable are grouped into one of two categories: nonmetallic or metallic.

Q29. If a cable is installed where it receives rough treatment, what should be added?

NONMETALLIC PROTECTION

The category of nonmetallic protective coverings is divided into three areas. These areas are (1) according to the material used as the covering, (2) according to the saturant in which the covering was impregnated, and (3) according to the external finish on the wire or cable. These three areas reflect three different methods of protecting the wire or cable. These methods allow some wire or cable to be classified under more than one category. Most of the time, however, the wire or cable will be classified based upon the material used as the covering regardless of whether or not a saturant or finish is applied.

Many types of nonmetallic materials are used to protect wires and cables. Fibrous braid is by far the most common and will be discussed first.

Fibrous Braid

Fibrous braid is used extensively as a protective covering for cables. This braid is woven over the insulation to form a continuous covering without joints (figure 1-15). The braid is generally saturated with asphalt, paint, or varnish to give added protection against moisture, flame, weathering, oil, or acid. Additionally, the outside braid is often given a finish of stearin pitch and mica flakes, paint, wax, lacquer, or varnish depending on the environment where the cable is to be used.

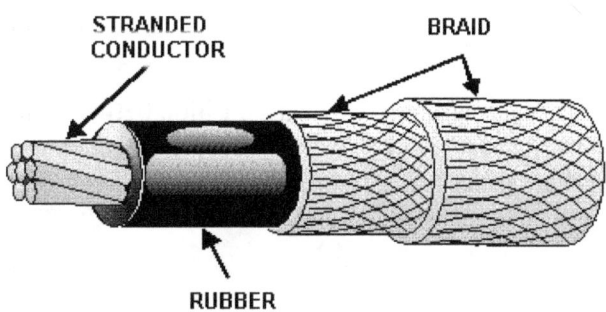

Figure 1-15.—Fibrous braid covering.

The most common type of fibrous braid is woven from light, standard, or heavy cotton yarn. Cotton yarn comes in different colors, which allows color-coding of the individual conductors. Cotton will not withstand all the possible environments in which a cable may be laid. Other materials currently being used to make fibrous braid are glazed cotton, seine twine or hawser cord, hemp, paper and cotton, jute, asbestos, silk, rayon, and fibrous glass. The choice of which material to use depends on the strength needed in the cable as well as how resistant it must be to its environment.

Fibrous Tape

Fibrous tape coverings are frequently used as a part of the protective covering of cables. The material of tape coverings is made into the tape before application to the cable. The material in yarns for braid covering is woven into fabric during the application to the cable. When tape covering is used, it is wrapped helically around the cable with each turn overlapping the previous turn.

The most common types of fibrous tape are rubber-filled cloth tape and a combination of cotton cloth and rubber. Except for duct tape, tape covering is never used as the outer covering on a cable. Tape coverings are used directly over the insulation of individual conductors and for the inner covering over the assembled conductors of a multiconductor cable. Frequently, tape coverings are used under the sheath of a lead-sheathed cable. Duct tape, which is made of heavy canvas webbing saturated with an asphalt compound, is often used over a lead-sheathed cable for protection against corrosion and mechanical injury.

Q30.   How many categories of nonmetallic protective coverings are there?

Q31.   What is the most common type of nonmetallic material used to protect wires and cables?

Q32.   What are the most common types of fibrous tape?

Woven Covers

Woven covers, commonly called loom, are used when exceptional abrasion-resistant qualities are required. These covers are composed of thick, heavy, long-fibered cotton yarns woven around the cable in a circular loom, much like that used on a fire hose. They are not braids, although braid covering are also woven; they are designated differently.

Rubber and Synthetic Coverings

Rubber and synthetic coverings are not standardized. Different manufactures have their own special compounds designated by individual trade names. These compounds are different from the rubber compounds used to insulate cable. These compounds have been perfected not for insulation qualities but for resistance to abrasion, moisture, oil, gasoline, acids, earth solutions, and alkalies. None of these coverings will provide protection against all types of exposure. Each covering has its own particular limitations and qualifications.

Jute and Asphalt Coverings

Jute and asphalt coverings are commonly used as a cushion between cable insulation and metallic armor. Frequently, they are also used as a corrosive-resistant covering over a lead sheath or metallic armor. Jute and asphalt coverings consist of asphalt-impregnated jute yarn heli-wrapped around the cable or of alternate layers of asphalt-impregnated jute yarn. These coverings serve as a weatherproofing.

Unspun Felted Cotton

Unspun felted cotton is commonly used only in special classes of service. It is made as a solid felted covering for a cable.

Q33.   What materials are commonly used as cushions between cable insulation and metallic armor?

METALLIC PROTECTION

Metallic protection is of two types: sheath or armor. As with all wires and cables, the type of protection needed will depend on the environment where the wire or cable will be used.

Metallic Sheath

Cables or wires that are continually subjected to water must be protected by a watertight cover. This watertight cover is either a continuous metal jacket or a rubber sheath molded around the cable.

Figure 1-16 is an example of a lead-sheathed (jacketed) cable used in power work. This cable is a standard three-conductor type. Each conductor is insulated and then wrapped in a layer of rubberized tape. The conductors are twisted together, and rope or fillers are added to form a round core. Over this is wrapped a second layer of tape called a serving. Finally, a lead sheath is molded around the cable.

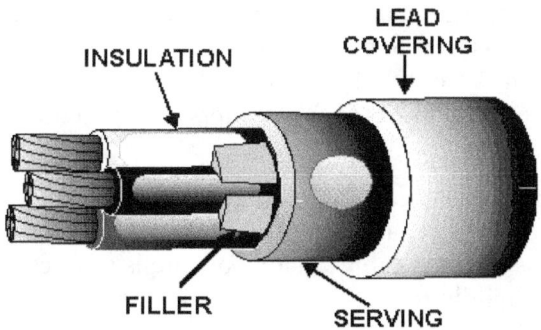

Figure 1-16.—Lead-sheathed cable.

Lead-sheathed cable is one of three types currently being used: alloy lead, pure lead, and reinforced lead. An alloy-lead sheath is much like a pure lead sheath but is manufactured with 2-percent tin. This alloy is more resistant to gouging and abrasion during and after installation. Reinforced lead sheath is used mainly for oil-filled cables where high internal pressures can be expected. Reinforced lead sheath consists of a double lead sheath. A thin tape of hard-drawn copper, bronze, or other elastic metal (preferably nonmagnetic) is wrapped around the inner sheath. This tape gives considerable additional strength and elasticity to the sheath, but must be protected from corrosion. For this reason, a second lead sheath is applied over the tape.

Metallic Armor

Metallic armor provides a tough protective covering for wires and cables. The type, thickness, and kind of metal used to make the armor depend on three factors: (1) the use of the conductors, (2) the environment where the conductors are to be used, and (3) the amount of rough treatment that is expected.

Figure 1-17 shows three examples of metallic armor cable: wire braid, steel tape, and wire armor.

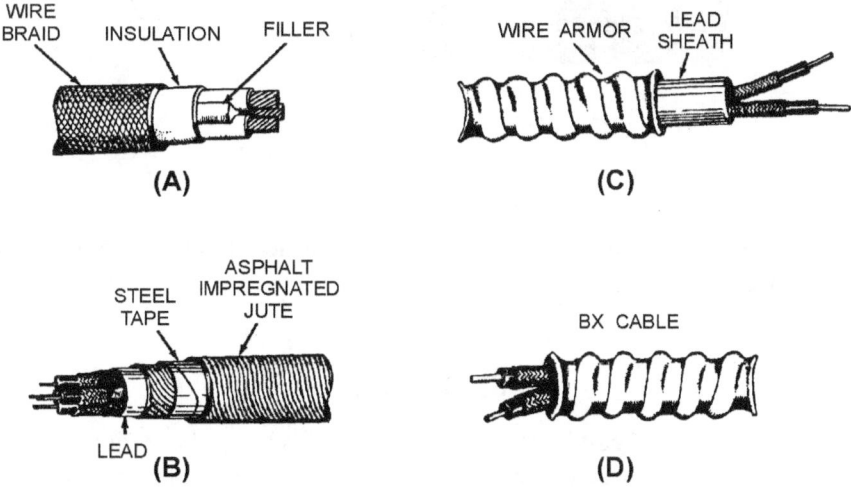

Figure 1-17.—Metallic armor cable.

WIRE-BRAID ARMOR.—Wire-braid armor (view A of figure 1-17), also known as basket-weave armor, is used when light and flexible protection is needed. Wire braid is constructed much like fibrous braid. The metal is woven directly over the cable as the outer covering. The metal used in this braid is galvanized steel, bronze, copper, or aluminum. Wire-braid armor is mainly for shipboard use.

STEEL TAPE.—A second type of metallic armor is steel tape. Steel tape covering (view B of figure 1-17) is wrapped around the cable and then covered with a serving of jute. There are two types of steel tape armor. The first is called interlocking armor. Interlocking armor is applied by wrapping the tape around the cable so that each turn is overlapped by the next and is locked in place. The second type is flat-band armor. Flat-band armor consists of two layers of steel tape. The first layer is wrapped around the cable but is not overlapped. The second layer is then wrapped around the cable covering the area that was not covered by the first layer.

WIRE ARMOR.—Wire armor is a layer of wound metal wire wrapped around the cable. Wire armor is usually made of galvanized steel and can be used over a lead sheath (see view C of figure 1-17). It can be used with the sheath as a buried cable where moisture is a concern, or without the sheath (view D of figure 1-17) when used in buildings.

Q34.   What are the two types of metallic protection?

Q35.   What are the three types of lead-sheathed cables?

Q36.   What are the three examples of metallic armor cable that were discussed?

COAXIAL CABLE

Coaxial cable (figure 1-18) is defined as two concentric wires, cylindrical in shape, separated by a dielectric of some type. One wire is the center conductor and the other is the outer conductor. These conductors are covered by a protective jacket. The protective jacket is then covered by an outer protective armor.

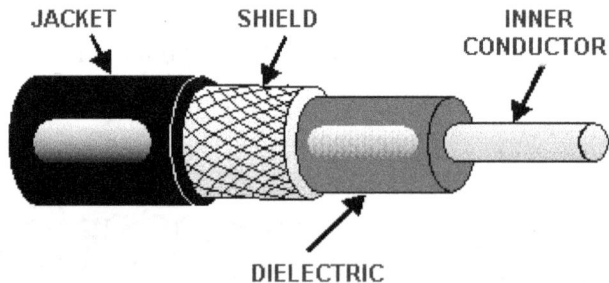

Figure 1-18.—Coaxial cable.

Coaxial cables are used as transmission lines and are constructed to provide protection against outside signal interference.

SUMMARY

In this chapter you learned that conductors are the means for tying the various components of an electrical or electronic system together. Many factors determine the type of conductor to be used in a specific application. In order for you to compare the different types and sizes of conductors, we discussed the following factors:

Unit Size—The unit size of a conductor is the mil-foot. A mil-foot is a circular conductor 1 foot long with a diameter of 1 mil (0.001 inch, or one-thousandth of an inch).

Conductor Sizes—The square mil and the circular mil are the units of measure used to determine the cross-sectional area of electrical conductors. The square mil, as it relates to a square conductor, is the cross-sectional area of a square conductor that has a side of 1 mil. The circular mil is the cross-sectional area of a circular conductor having a diameter of 1 mil. The circular mil area (CMA) of a conductor is computed by squaring the diameter of the circular conductor is mils. Thus, a wire having a diameter of 4 mils (0.004 inch) has a CMA of 4^2, or 16 circular mils. If the conductor is stranded, the CMA for a strand is computed, and the CMA for the conductor is computed by multiplying the CMA of the strand by the number of strands. The relationship of the square mil to the circular mil is determined by comparing the square mil area of a circular conductor having a diameter of 1 mil ($A = \pi r^2$) to the circular mil area of the same conductor (D^2). Therefore, there is 0.7854 square mil to 1 circular mil. There are more circular mils than square mils in a given area.

Specific Resistance—The specific resistance of a substance is the resistance in ohms offered by a unit volume (the circular-mil-foot) to the flow of electric current. The three factors that are used to calculate the specific resistance of a particular conductor are (1) its length, (2) its cross-sectional area, and (3) the specific resistance of a unit volume of the substance from which the conductor is made. The specific resistance for various sizes and lengths of standard solid copper wire can be determined by the use of tables.

Wire Gauge—A wire gauge is used to determine the American Standard Wire Gauge size of conductors. The measurement of a bare conductor is taken in the slot, not in the circular area at the bottom of the slot.

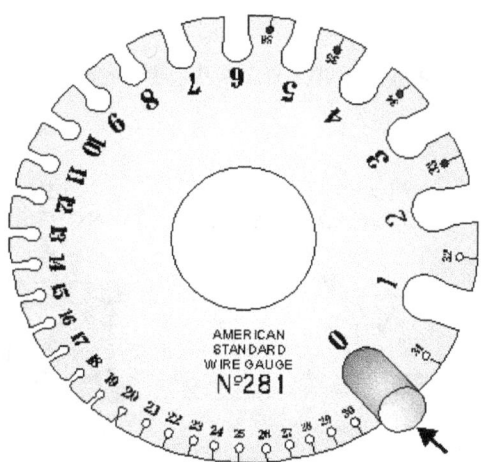

Selection of Wire Size—Four factors must be considered in selecting the proper wire size for a particular electrical circuit. These factors are (1) conductor size, (2) the material it's made of, (3) the location of the wire in the circuit, and (4) the type of insulation used. Some of the types of insulation used in a high-temperature environment are FEP, extruded polytetrafluoroethylene, and silicone rubber. The ambient (surrounding) temperature of a conductor is an important part of total conductor heating.

Copper-versus-Aluminum Conductors—The two most common metals used for electrical conductors are copper and aluminum. Some advantages of copper over aluminum as a conductor are that copper has higher conductivity, is more ductile, has a higher tensile strength, and can be easily soldered. Two advantages of aluminum wire for carrying electricity over long distances are its lightness and it reduces corona (the discharge of electricity from a wire at high potential).

Temperature Coefficient of Resistance—The temperature coefficient of resistance is the amount of increase in the resistance of a 1-ohm sample of a conductor per degree of temperature rise above 0° C. The resistance of copper and other pure metals increases with an increase in temperature.

Conductor Insulation—Insulators have a resistance that is so great that, for all practical purposes, they are nonconductors. Two fundamental properties of insulating materials are (1) insulation resistance and (2) the resistance to current leakage through the insulation. Dielectric strength is the ability of the insulation material to withstand potential difference. The dielectric strength of an insulator is determined by raising the voltage on a test sample until it breaks down.

Insulating Materials—Some common insulating materials have properties and safety precautions that should be remembered. These are:

- The purpose of coating a copper conductor with tin when rubber insulation is used is to prevent the insulation from deteriorating due to chemical action.

- When extruded polytetrafluoroethylene insulation is heated, caution should be observed not to breathe the vapors.

- The most commonly used insulating materials for extremely high-voltage conductors are varnished cambric and oil-impregnated paper.

- Magnet wire is the common name for enamel-insulated wire used in meters, relays, small transformers, motor windings, and so forth.

- The Navy is getting away from using asbestos insulation because asbestos fibers can cause lung disease and/or cancer.

- Asbestos insulation becomes a conductor when it gets wet.

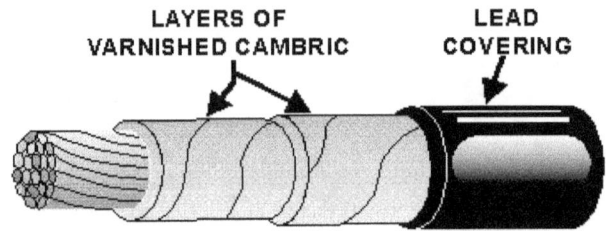

LAYERS OF VARNISHED CAMBRIC   LEAD COVERING

Conductor Protection—There are several types of conductor protection in use. The type commonly used aboard Navy ships is wire-braid armor.

ANSWERS TO QUESTIONS Q1. THROUGH Q36.

A1.   *To allow comparisons between conductors of different sizes and resistance.*

A2.   *375 mils (move the decimal three places to the right).*

A3.   *A circular conductor with a diameter of 1 mil and a length of 1 foot.*

A4.   *The cross-sectional area of a square conductor with a side of 1 mil.*

A5.   *The cross-sectional area of a circular conductor with a diameter of 1 mil.*

A6.   *Circular mil area (CMA) = D^2 (in mils) $\times$ number of strands 0.0004 inch = 4 mils (CMA) = $4^2 \times$ 19 (strands)(CMA) = 16×19 = 304 mils.*

A7.   *The resistance of a unit volume of a substance.*

A8.   *Length, cross-sectional area, and specific resistance of a unit volume of the substance from which the conductor is made.*

A9.   *1,000 ft = 10.4 ohms 1,500 ft = 1.5×0.4 = 15.6 ohms*

A10.   *In the parallel walled slot not the circular area.*

A11.   *Conductor size, the material it is made of the location of the wire in a circuit, and the type of insulation used.*

A12.   *FEP, extruded polytetrafluoroethylene, and silicone rubber.*

A13.   *The heat surrounding the conductor is an important part of total conductor heating.*

A14.   *It is light and reduces corona.*

A15.   *It has higher conductivity, it is more ductile, it has relatively high tensile strength, and it can be easily soldered.*

A16.   *The amount of increase in the resistance of a 1-ohm sample of the conductor per degree of temperature rise above $0°$ C*

A17.   *It increases.*

A18.   *Conductors have a very low resistance and insulators have a resistance that is so great that, for all practical purposes, they are nonconductors.*

A19.   *Insulation resistance and dielectric strength.*

A20.   *The resistance to current leakage through the insulation.*

A21.   *The ability of the insulation material to withstand potential difference.*

A22.   *By raising the voltage on a test sample until it breaks down.*

A23.   *To prevent the rubber insulation from deteriorating due to chemical action.*

A24.   *Avoid breathing the vapors when the insulation is heated.*

A25.   *Breathing asbestos fibers can cause lung disease and/or cancer*

A26.   *It will become a conductor.*

A27.   *Varnished cambric and oil-impregnated paper.*

A28.   *Magnet wire.*

A29.   *Metallic coat.*

A30.   *Three.*

A31.   *Fibrous Braid.*

A32.   *Rubber-filled cloth tape and a combination of cotton cloth and rubber.*

A33.   *Jute and Asphalt coverings.*

A34.   *Sheath and armor*

A35.   *Alloy lead, pure lead, and reinforced lead.*

A36.   *Wire braid, steel tape, and wire armor*

www.ingramcontent.com/pod-product-compliance
Lightning Source LLC
Chambersburg PA
CBHW080626180526
45168CB00007B/3068